INSTRUCTIONS FOR

HOW TO REHABILITATE A RETURNING SOLDIER

ILLUSTRATED BY
SIGN-AL

GULF WAR EDITION

STACEY INTERNATIONAL

INSTRUCTIONS FOR RE-USE

Don't be surprised if he reacts violently to men with moustaches – he has learned to mistrust them.

INSTRUCTIONS FOR RE-USE

Discourage family visits for the first two weeks – he has a lot of pent-up frustration to get rid of.

INSTRUCTIONS FOR RE-USE

It may take a while to discourage your spouse from smoking – be patient, the Chancellor of the Exchequer will cure her.

INSTRUCTIONS FOR RE-USE

Don't be alarmed if he drives at night without headlights or indicators – he can see where he's going.

INSTRUCTIONS FOR RE-USE

**Be reassuring if he runs for cover
when the car alarm goes off.
(Be less forgiving if he leaves you behind.)**

INSTRUCTIONS FOR RE-USE

His desire to tear down statuary may
seem irrational. He has been encouraged
to do so recently, but with patience
can be cured.

INSTRUCTIONS FOR RE-USE

Don't take it personally if your partner goes to bed at sunset, fully clothed, and reads by torchlight.

INSTRUCTIONS FOR RE-USE

Warn the neighbours that he may shower in the back garden (but only every three days). Put your foot down when he washes his clothes in the remaining water.

INSTRUCTIONS FOR RE-USE

Don't worry if he tries to dismantle the bed every morning.

INSTRUCTIONS FOR RE-USE

Don't be surprised if he sets fire to the dustbin to get hot water for a shave. Keep the matches hidden.

INSTRUCTIONS FOR RE-USE

**.... Keep those matches hidden!
He may try to brew up in the bedroom.**

INSTRUCTIONS FOR RE-USE

Comfort him during bouts of depression when there is no mail for him.

INSTRUCTIONS FOR RE-USE

He may be shocked at first to see you without a veil, but he'll soon get used to it.

Ignore him if he stands outside the kitchen window waiting for grub – hunger will eventually drive him in.

Don't be surprised if he tries to flag down passing tankers.

INSTRUCTIONS FOR RE-USE

Do not, on any account, suggest going on a camping holiday.

INSTRUCTIONS FOR RE-USE

Humour him if he tries to chat up members of the opposite sex – even relatives.

Discourage him from throwing the cutlery and crockery in the bin after meals.